Kijc

WHAT DO CONSTRUCTION WORKERS DO ALL DAY?

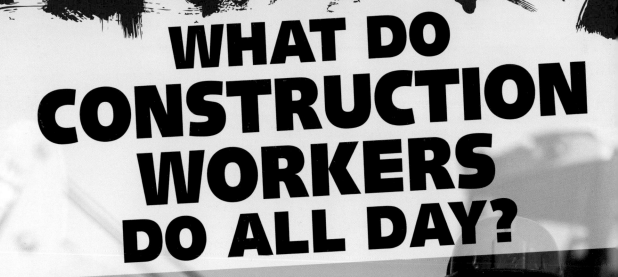

By Emily Mahoney

Gareth Stevens
PUBLISHING

Please visit our website, www.garethstevens.com. For a free color catalog of all our high-quality books, call toll free 1-800-542-2595 or fax 1-877-542-2596.

Library of Congress Cataloging-in-Publication Data

Names: Mahoney, Emily Jankowski, author.
Title: What do construction workers do all day? / Emily Mahoney.
Description: New York : Gareth Stevens Publishing, [2021] | Series: What do they do? | Includes index.
Identifiers: LCCN 2019050750 | ISBN 9781538256855 (library binding) | ISBN 9781538256831 (paperback) | ISBN 9781538256848 (6 Pack) | ISBN 9781538256862 (ebook)
Subjects: LCSH: Construction workers—Juvenile literature.
Classification: LCC HD8039.B89 M34 2021 | DDC 624.023—dc23
LC record available at https://lccn.loc.gov/2019050750

Published in 2021 by
Gareth Stevens Publishing
111 East 14th Street, Suite 349
New York, NY 10003

Editor: Emily Mahoney
Designer: Laura Bowen

Photo credits: Series art Dima Polies/Shutterstock.com; cover, p. 1 FotoAndalucia/Shutterstock.com; p. 5 Xinhua News Agency/Contributor/Xinhua News Agency/Getty Images; p. 7 Andersen Ross/The Image Bank/Getty Images Plus/Getty Images; p. 9 xavierarnau/E+/Getty Images; p. 11 LivingImages/E+/Getty Images; p. 13 FatCamera/E+/Getty Images; p. 15 Joseph Sohm/Corbis NX/Getty Images Plus/Getty Images; p. 17 Adina Tovy/robertharding/Getty Images Plus; p. 19 Jeff Greenberg/Contributor/Universal Images Group/Getty Images; p. 21 kali9/E+/Getty Images.

Printed in the United States of America

Some of the images in this book illustrate individuals who are models. The depictions do not imply actual situations or events.

CPSIA compliance information: Batch #CS20GS: For further information contact Gareth Stevens, New York, New York, at 1-800-542-2595.

Find us on

CONTENTS

Boldface words appear in the glossary.

An Important Job

Any building that you have ever been in, or any street that you have walked on, wouldn't exist without construction workers! A construction worker's day is fun, but it's a lot of hard work. It can also be dangerous, or unsafe. Keep reading to see what construction workers do all day!

Prep Work

The first thing that a construction worker must do is learn about the project they are working on. The **site supervisor** meets with the workers to tell them about the job. They also tell each worker what they are **responsible** for completing. Each member of the team must complete their part of the job.

Starting a Job

When a construction worker arrives at the construction site, there's a lot to do! Sometimes the site needs to be cleared of **debris** or garbage before they can start working. Once the site is cleared, all of the building **materials** must be unloaded, usually from a supply truck.

Once the site is prepared, the building can begin! Construction workers might do jobs such as pouring concrete. They set up **scaffolding** to help with building. They help painters, electricians, or others with their jobs. They may even help to put up walls for a building!

11

Working on Roads

Some construction workers are responsible for fixing or making roads. In this case, the workers sometimes need to direct **traffic** or set up traffic signs before they begin. This is to make sure that everyone at the site and the people driving their cars are safe.

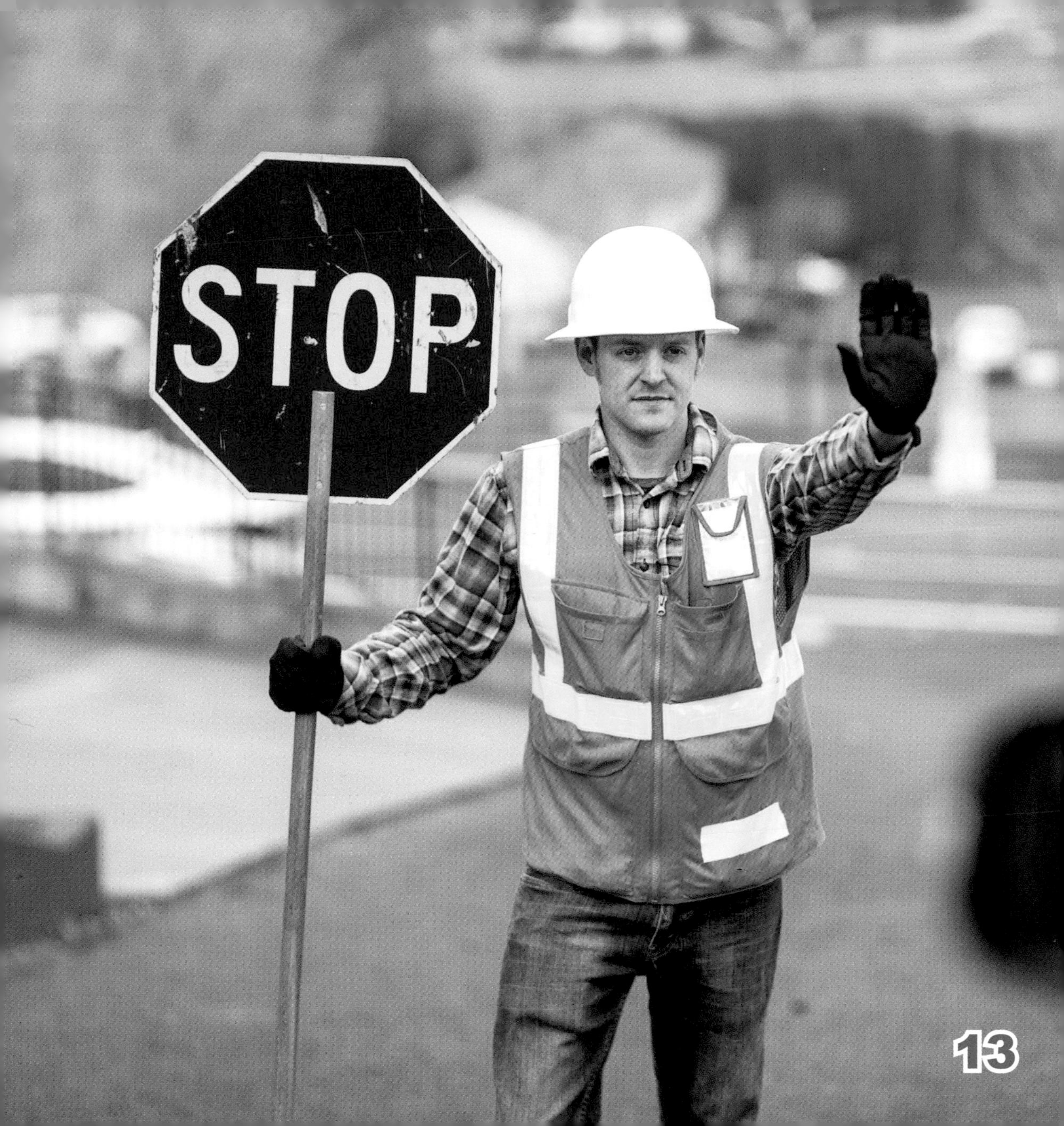
STOP

Working on roads is a busy job. Some things that construction workers might do are digging space for new roads. They help **operate** machines and may lay asphalt, or the matter the road is made of. They set up concrete **barriers** or help to paint lines on the new roads.

15

Building Bridges

Sometimes road work involves making or fixing bridges. In that case, workers who **specialize** in different areas, such as concrete, might be brought in to help the construction workers. It's especially important to make sure that the bridge can hold the weight of the cars driving over it!

Cleaning Up

Once the project is finished, there's still work to be done. Construction workers are responsible for cleaning up their work site. They make sure everything is safe before people begin using the new road or building. They want the project to look its best since they worked hard on it!

19

The Finished Product

A construction worker's job may be hard, but it feels good when the work is done. From fixing roads to creating a building, each day in the life of a construction worker might look different. But, working with others to finish a project is a fun job!

GLOSSARY

barrier: something that makes progress hard

debris: the remains of something that has been broken

material: matter from which something is made

operate: to use and control something

responsible: having the job or duty of dealing with or taking care of something or someone

scaffolding: metal poles and wooden boards used to make a structure used during construction

site supervisor: someone who manages construction workers at a job site

specialize: to be good at one specific thing

traffic: all the vehicles, such as cars and trucks, moving along a roadway

FOR MORE INFORMATION

BOOKS

Ng, Yvonne. *They're Tearing Up Mulberry Street.* Mankato, MN: Amicus, 2021.

Schuh, Mari C. *All About Construction Workers.* Minneapolis, MN: Lerner Publications, 2021.

WEBSITES

Architecture, Building & Construction Links for Kids
www.b4ubuild.com/kids/kidlinks.html
This site has interesting information about some of the most impressive structures ever built, like the Space Needle.

Engineering Games
pbskids.org/games/engineering/
You can learn about engineering and play fun games on this PBS website.

INDEX